Good Germs In My Gut

The gut microbiome for readers of all ages

Rene Chee, PhD

To Bennett,

whose love of reading, learning and eating inspired this book.

An Act On Science Book

Illustrations and cover design by Rene Chee
Bacteria caricatures are based on microscope images of the bacteria mentioned on pages 10, 14-19.

ISBN 978-0-9978200-2-7

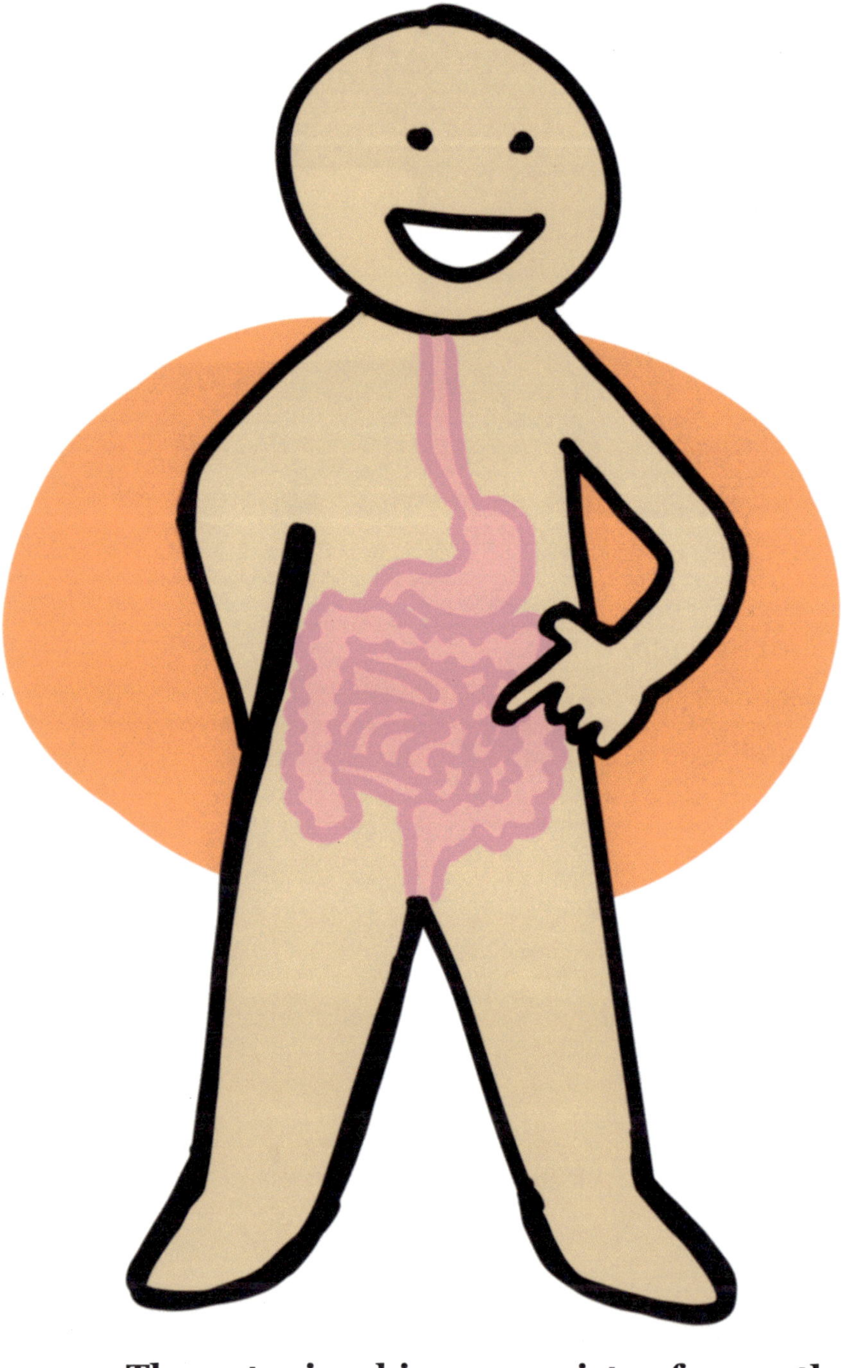

This is my gut.

The gut microbiome consists of more than 100 trillion bacteria in the small and large intestines.

Gut bacteria are naturally shed in feces and make up 55% of the dry weight of feces.

There are
many germs in my gut.

Germs are
very
very
small.

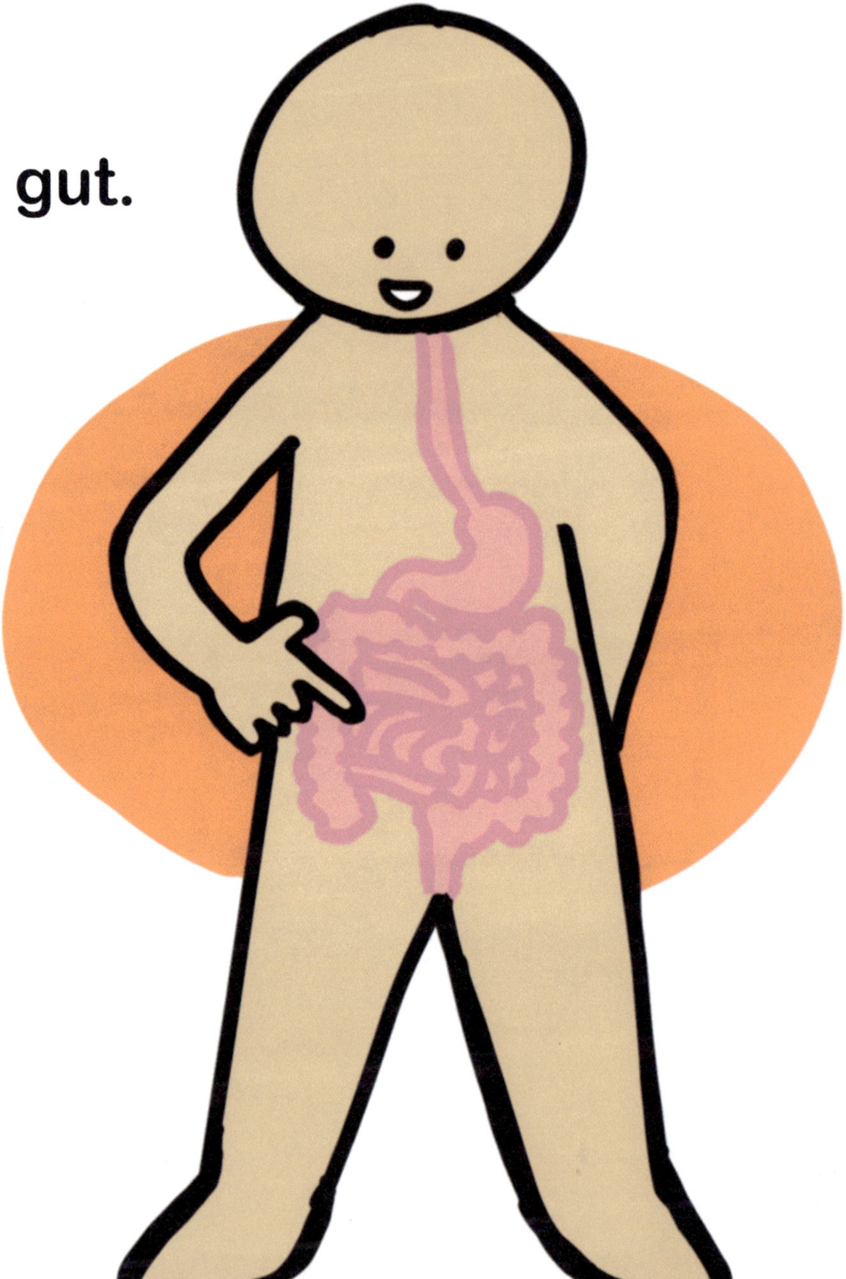

Our health is influenced by our genes, which are hard to change. However, our health is also heavily influenced by our gut bacteria, which we can affect.

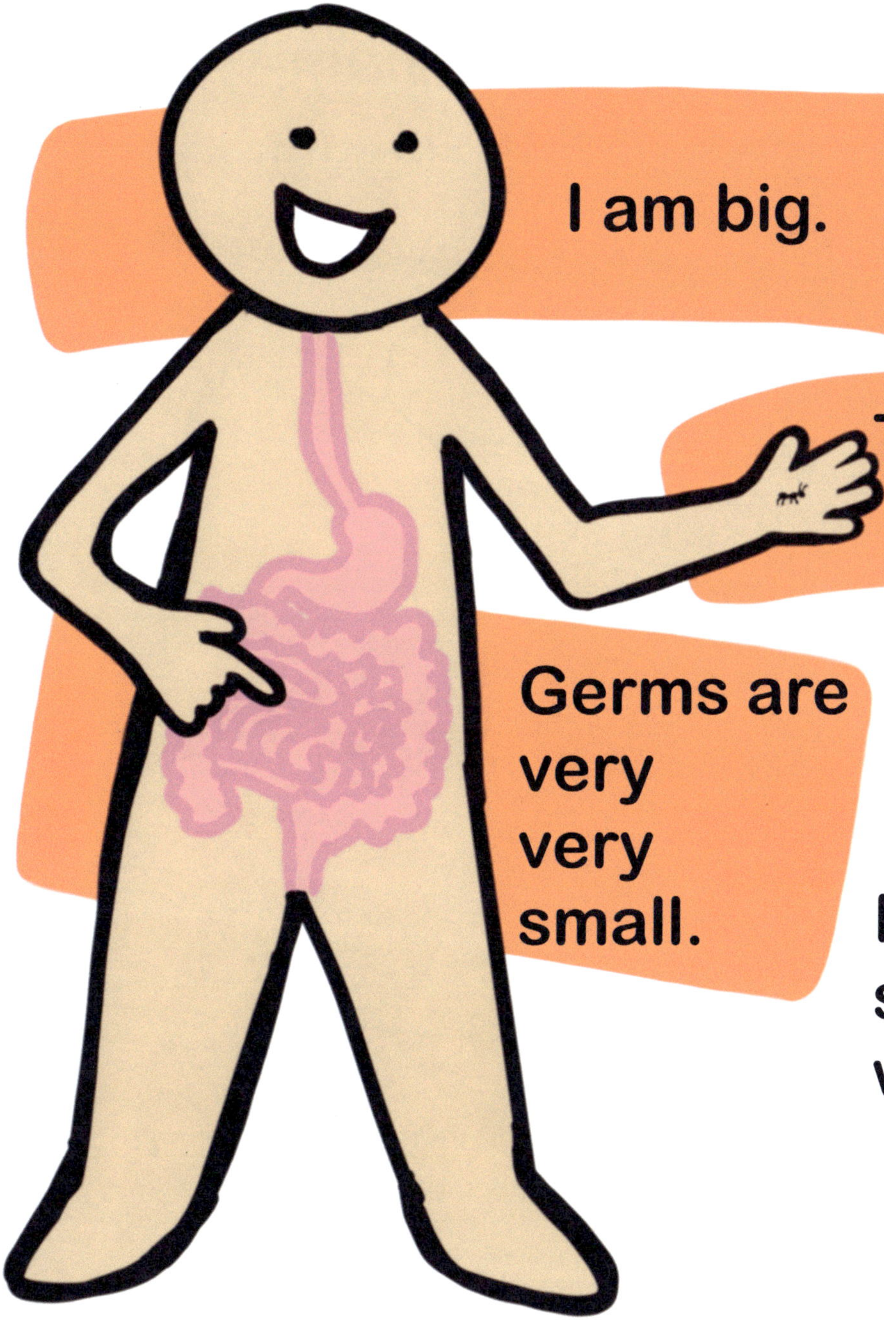

I am big.

The ant is small.

I can see the ant.

Germs are very very very small.

I cannot see germs with my eye.

I need help
to see germs.

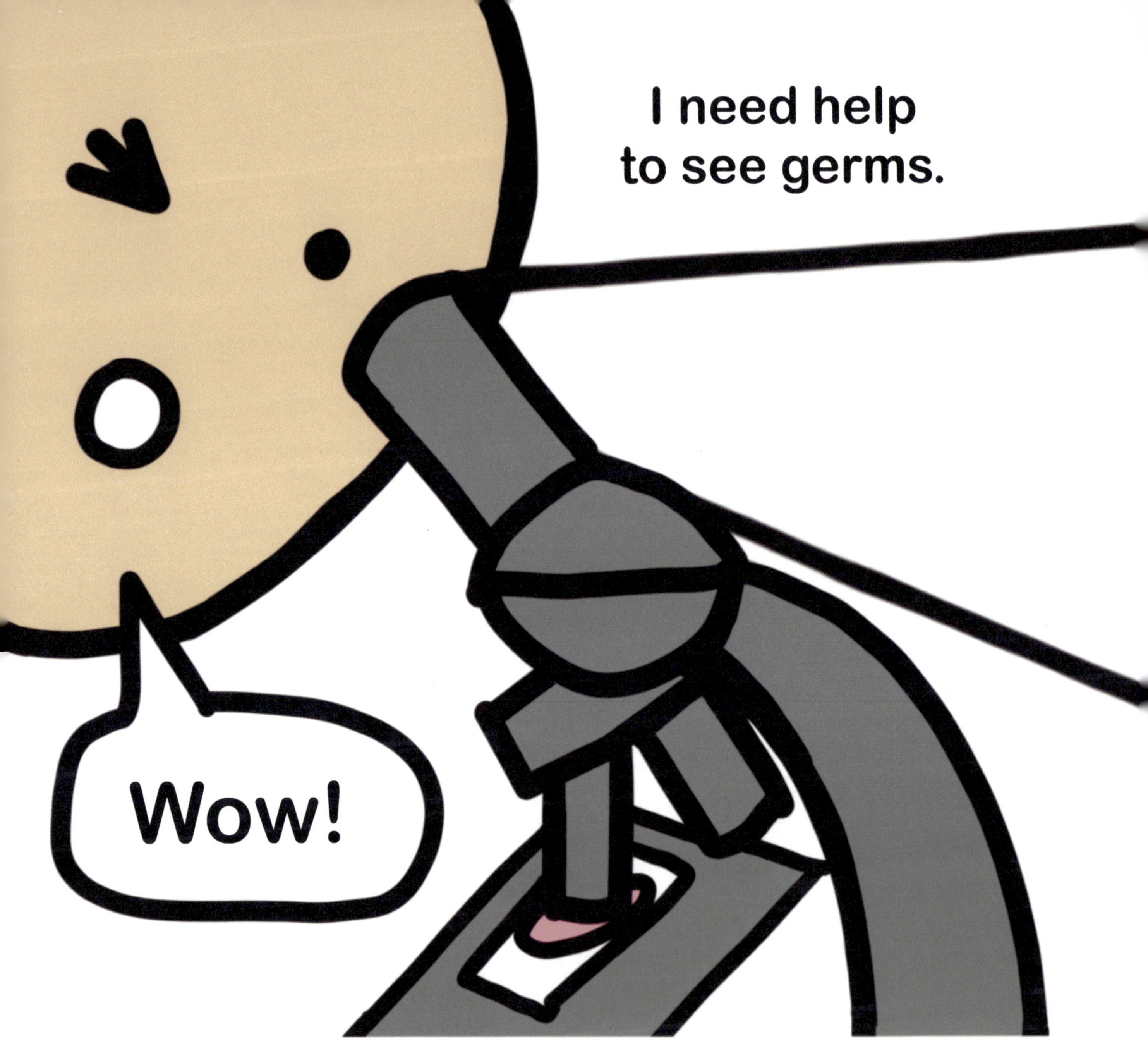

Gut bacteria can be seen with a light microscope.

Now I can see the germs!

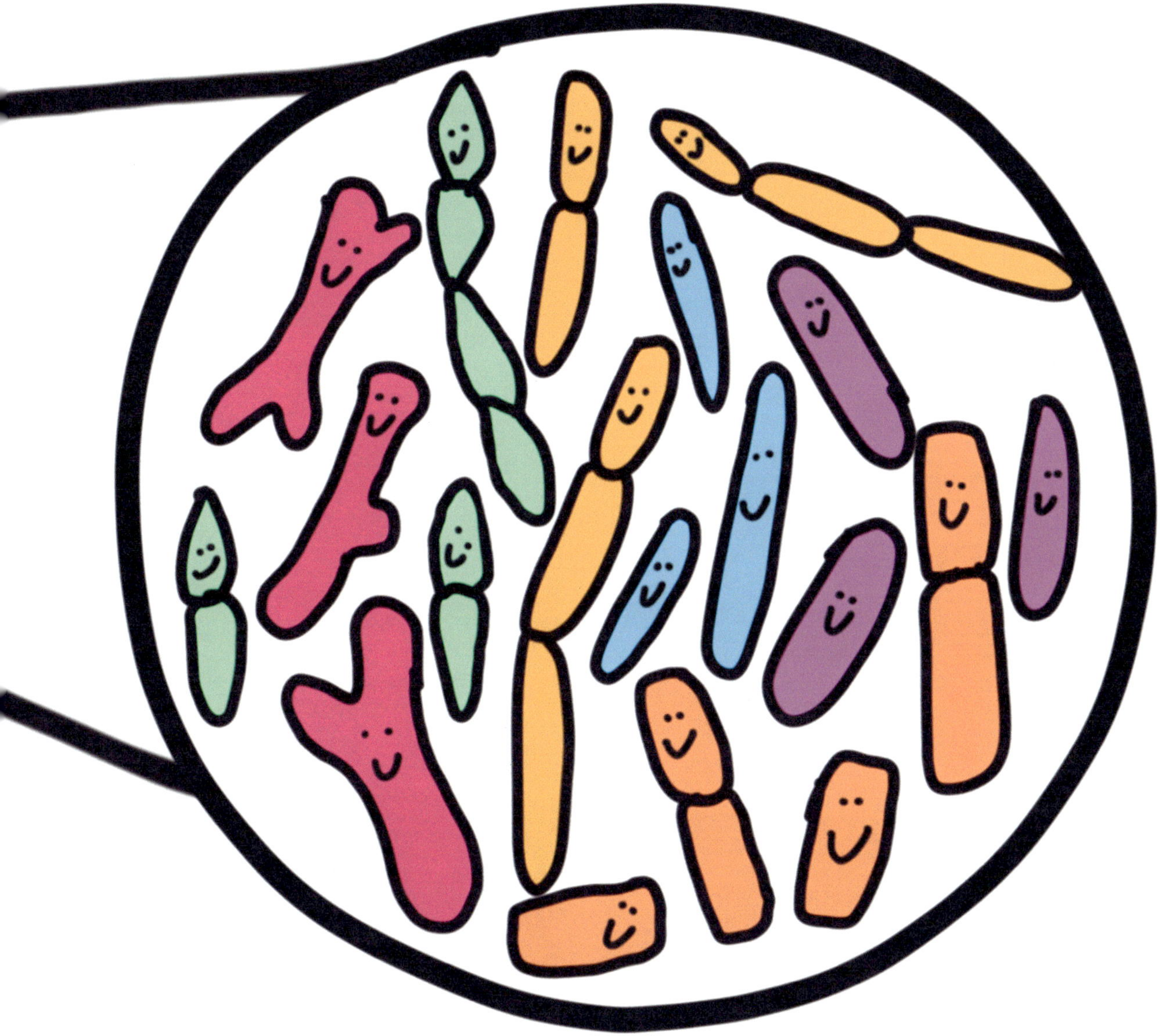

The composition of bacteria in our gut can give us clues about our health. We can find out the composition by collecting fecal samples and sending them to a lab for analysis.

There are
bad germs and good germs.

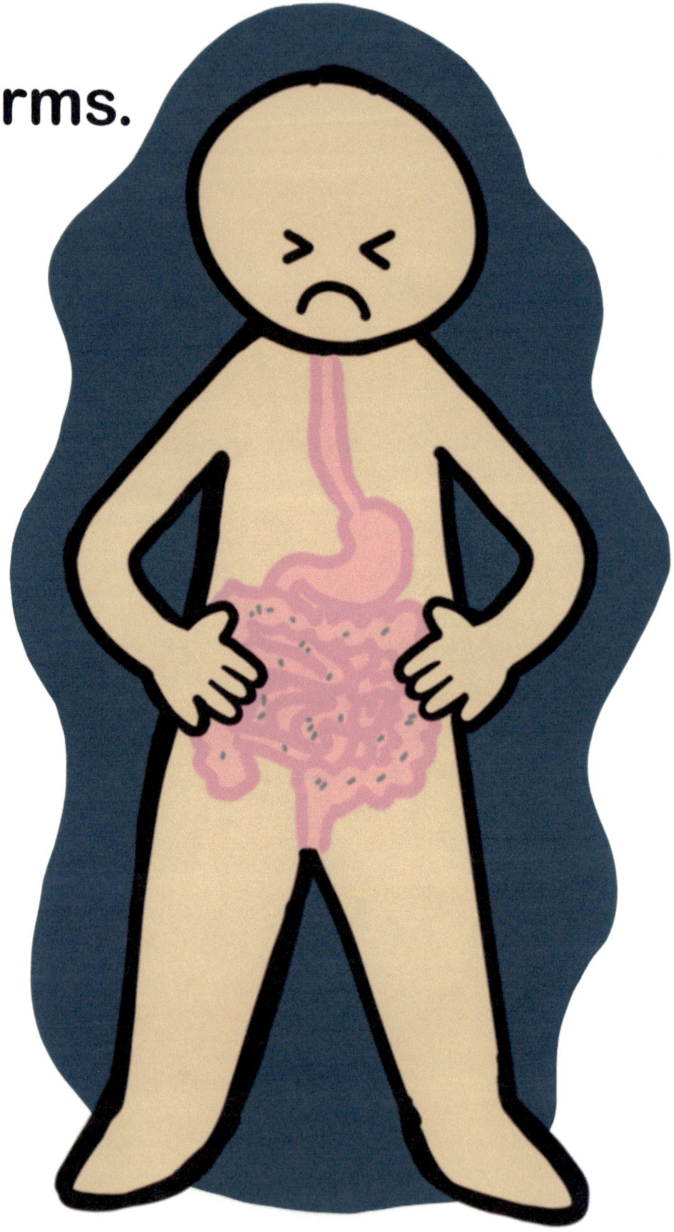

Bad germs in my gut
make me sick.

Pathogenic bacteria, like *Listeria* or *Salmonella*, can cause food poisoning.
In response, our immune system tries to eliminate them,
manifesting as diarrhea or fever.
If the immune system fails to keep the bacteria in check,
the result can be life-threatening.

Good germs in my gut
make me happy and strong!

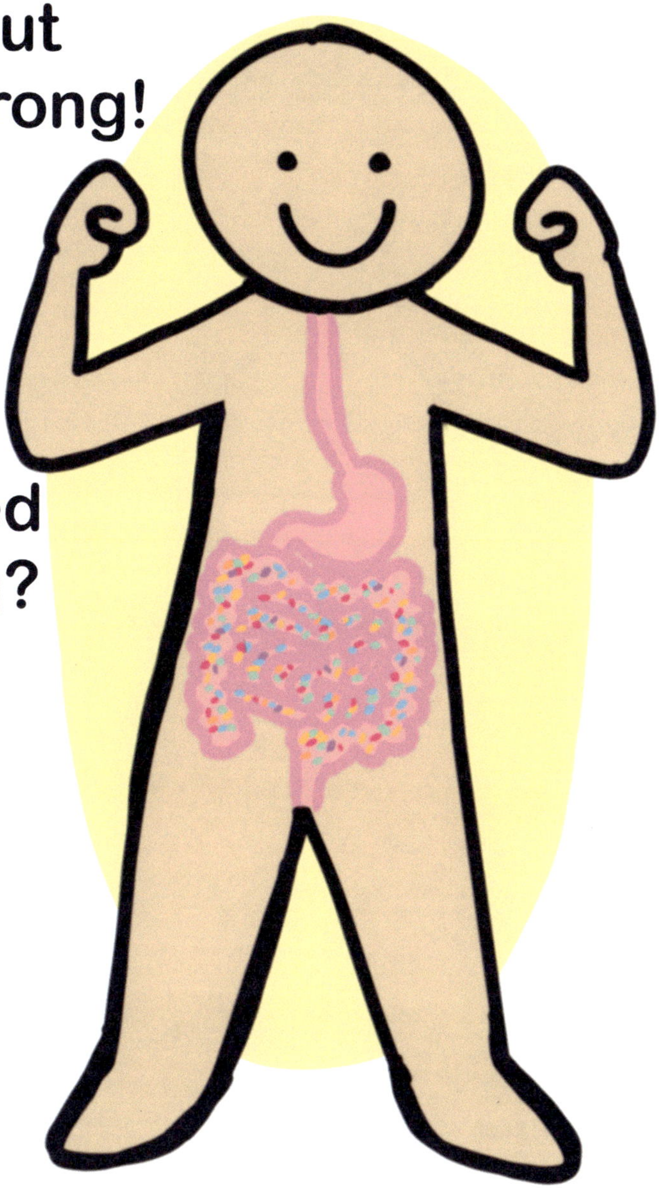

How many kinds
of good germs do I need
to be happy and strong?

One kind? No.

Two kinds? No.

Many kinds? Yes!

A diverse gut microbiome is associated with a range of health benefits, including a stronger immune system, decreased obesity, decreased anxiety and a lower risk of chronic conditions such as diabetes, heart disease and cancer.

When good germs fill my gut, there is no room for bad germs.

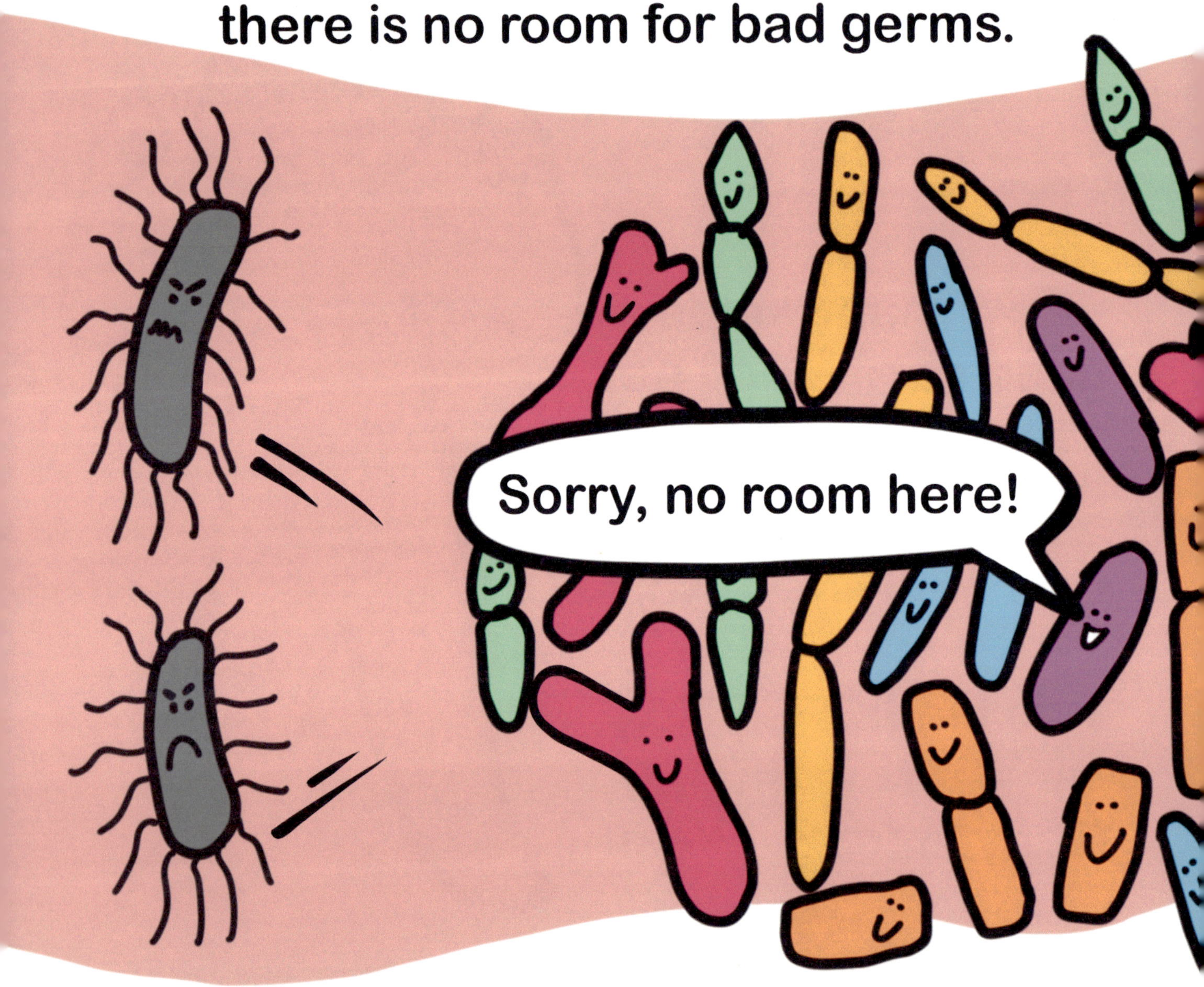

After taking antibiotics, which can wipe out the gut microbiome, the chance of a life-threatening *C. difficile* bacterial infection increases by 7-10 times.

Good germs make me happy and strong.

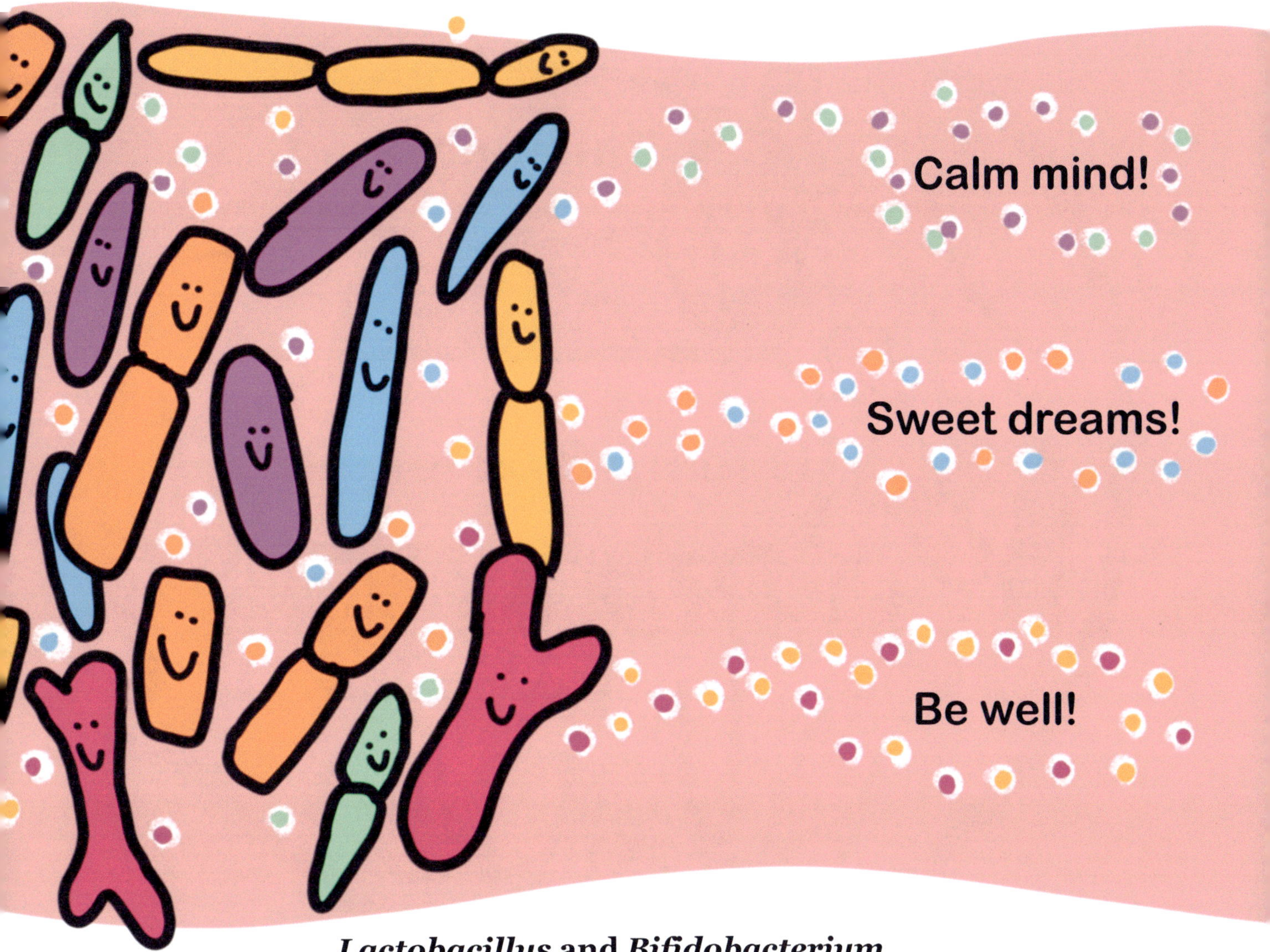

Lactobacillus **and** ***Bifidobacterium***
decrease stress, depression and anxiety.

Butyrate, a compound made by beneficial gut bacteria, has been shown to improve sleep, strengthen the immune system, control blood sugar levels, and reduce inflammation in the gut.

Good germs need to eat.

**Prebiotics are foods that gut bacteria feed on.
They are mainly soluble fibers found in
whole grains, beans, nuts, fruits and vegetables.**

What do you want to eat?

Apples encourage the growth of the anti-inflammatory bacteria *Faecalibacterium prausnitzii.*

Eating broccoli increases *Bacteroides* bacteria in the gut. *Bacteroides* bacteria break down broccoli into the cancer-fighting compound sulforaphane.

Oatmeal consumption increases *Blautia* bacteria in the gut which is correlated with improved cholesterol levels.

Eating nuts increases *Clostridium*, *Lachnospira* and *Roseburia* bacteria, all of which are known to make butyrate.

Consuming a cup of beans every day for 8 weeks improves the gut microbiome composition and decreases LDL cholesterol.

Taking a high amount of flavanol (found in dark chocolate) for 4 weeks increased *Bifidobacterium* and *Lactobacilli* in the gut.
This was associated with lower levels of inflammation and cholesterol.

I feed my good germs
when I eat the foods they like.

How can I get more good germs?

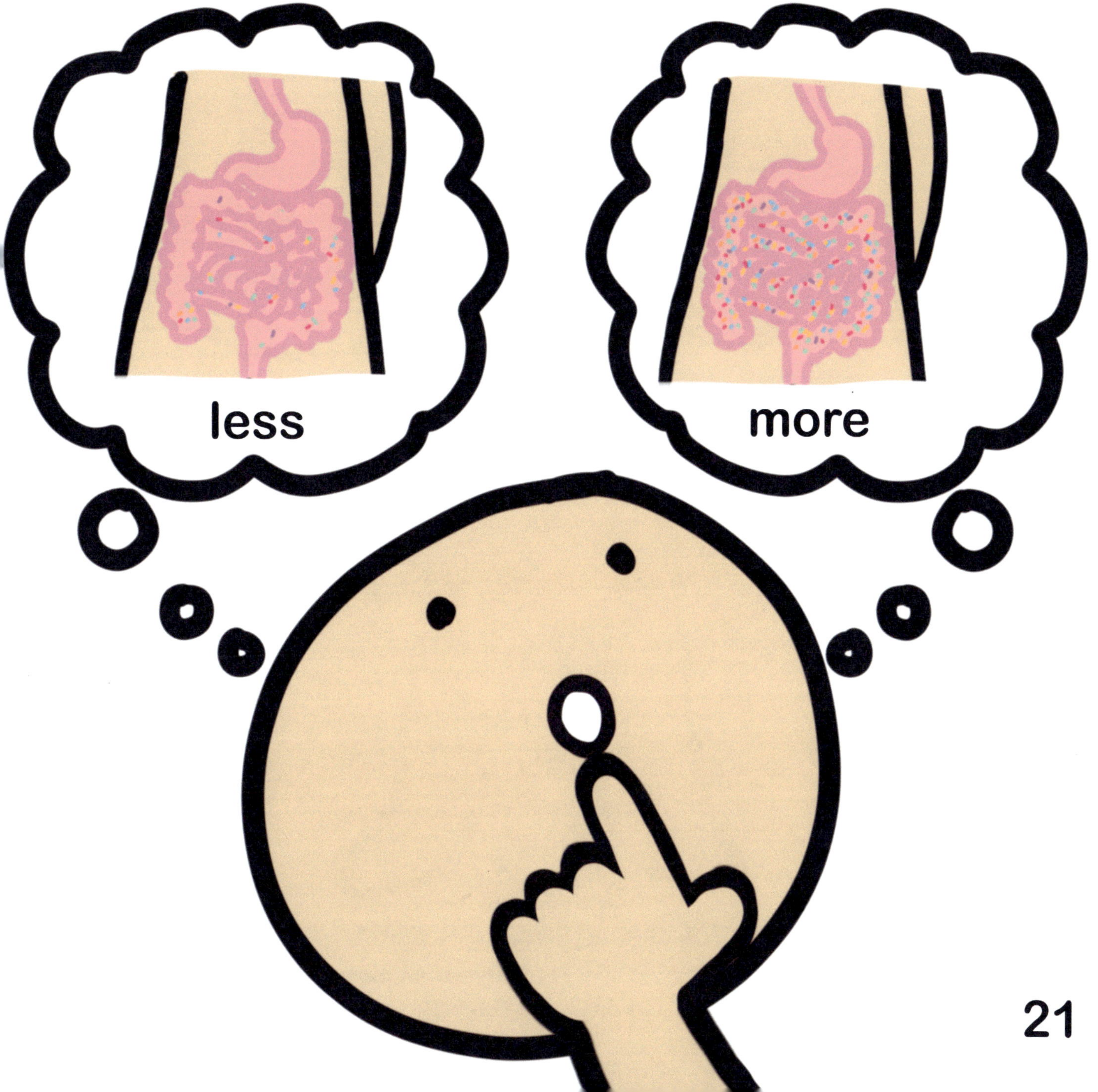

I can eat foods that have good germs.

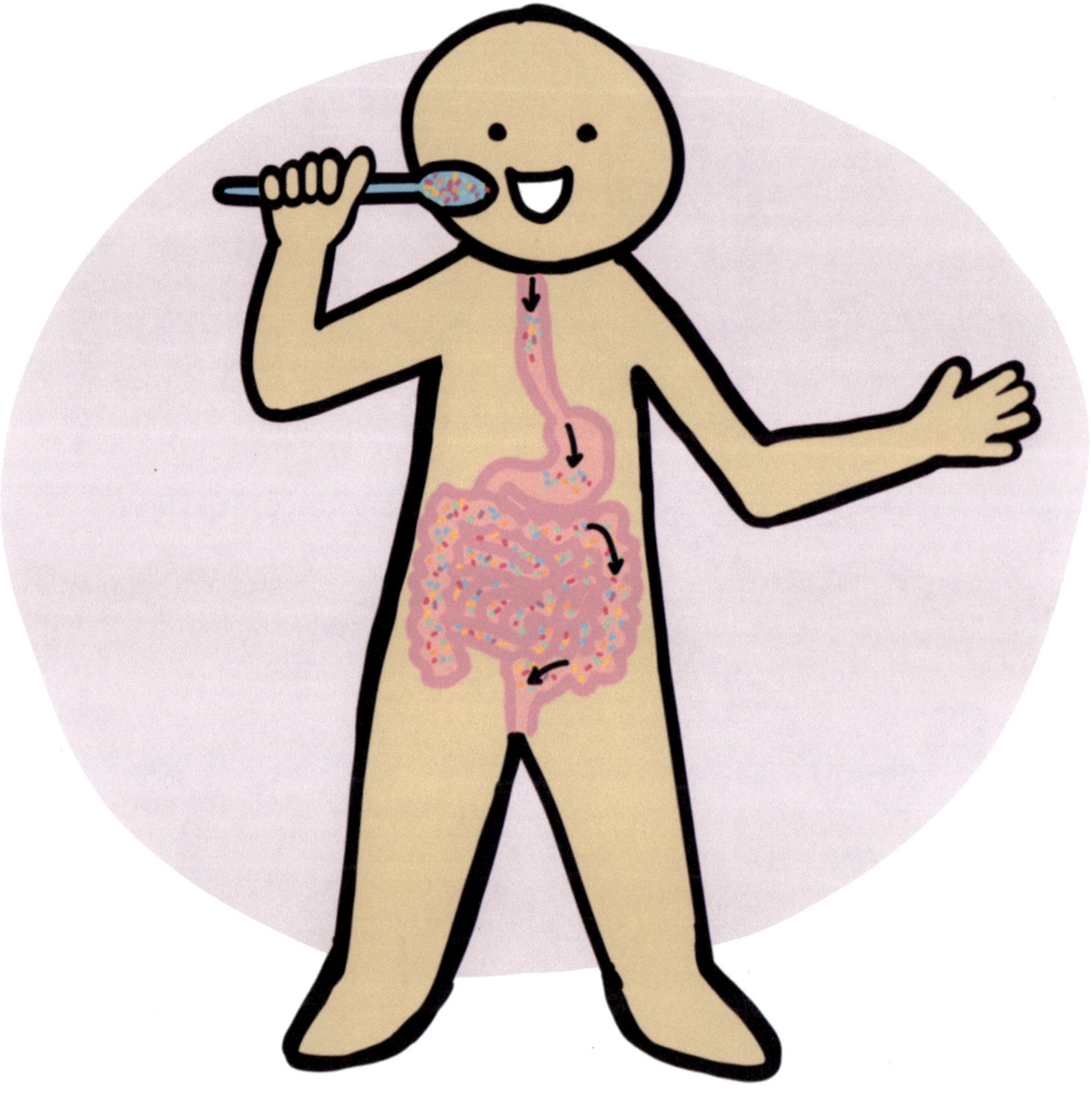

Fermented foods have live and active bacteria.

Eating fermented foods for 10 weeks increases gut microbiome diversity and decreases inflammation.

What foods have good germs?

These foods are sour, salty and have good germs!

Fermented vegetables like sauerkraut and kimchi have many more kinds of bacteria compared to probiotic pills.

One study found 28 different strains of bacteria in sauerkraut.

Eating kimchi is associated with lower rates of allergies, asthma and eczema.

These foods are cold, creamy and have good germs!

Yogurt and kefir are fermented milk products.

Eating yogurt every day lowers the risk of heart disease and type 2 diabetes.

Diabetic patients taking 2 cups of kefir daily for 8 weeks lowered their blood glucose levels.

This food is sticky, stringy and has good germs!

Natto is made by fermenting soybeans with *Bacillus subtilis*.

In natto, *B. subtilis* make large amounts of vitamin K2. Vitamin K2 helps cardiovascular and bone health by directing calcium from the lining of blood vessels into bones.

***B. subtilis* also increase interferon gamma, an anti-inflammatory molecule that protects against viral infections.**

I eat foods with good germs and I eat foods that good germs eat.

Eating probiotics (live good bacteria) together with prebiotics (food that good bacteria eat) can increase the number of good bacteria in your gut because the probiotics can grow and multiply on the prebiotics.

I have many good germs in my gut.
My good germs are full and happy!

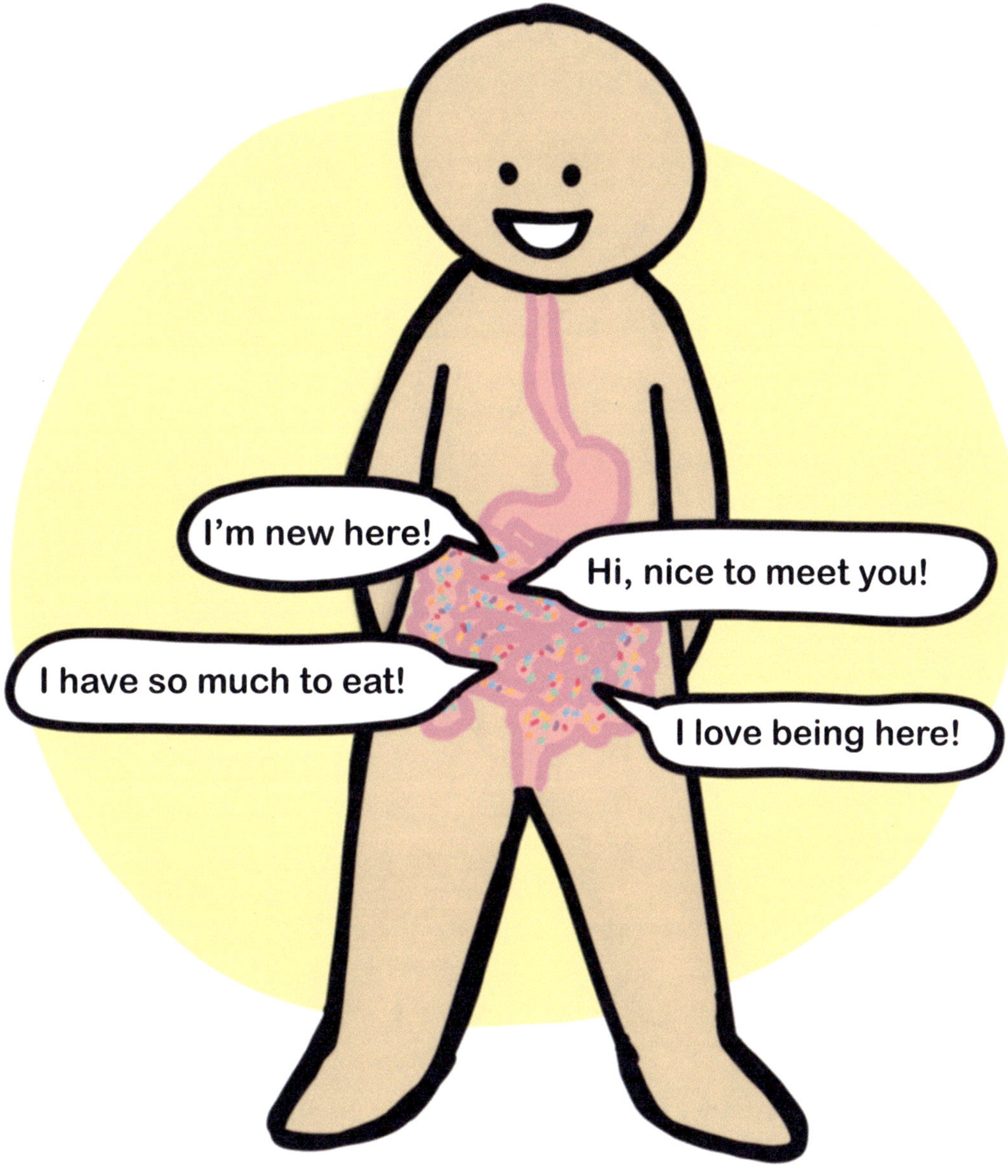

Many happy good germs in my gut make me happy and strong!

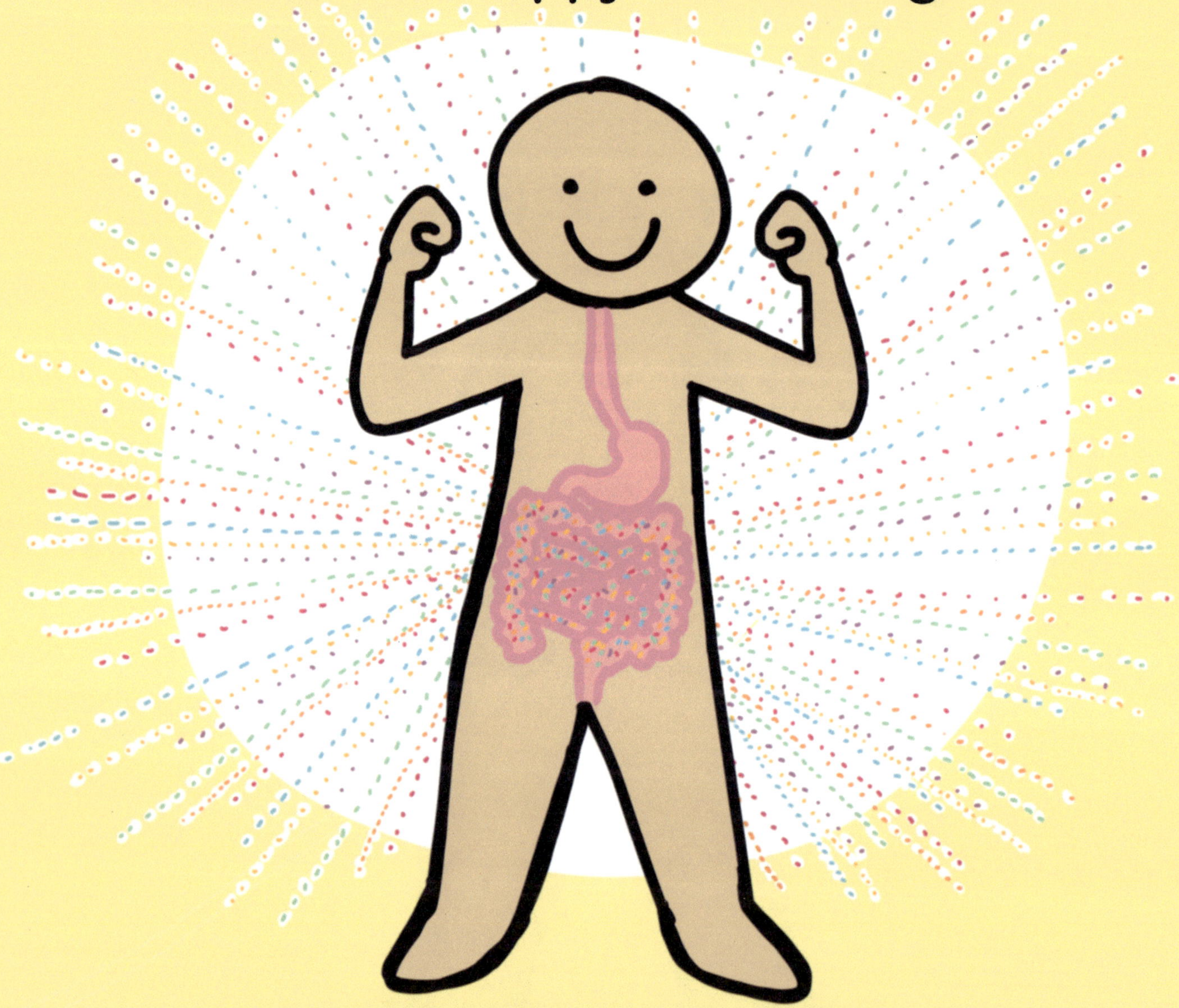

When gut bacteria are actively feeding,
they make vitamins, enzymes and compounds
that benefit the brain, immune system and metabolism.

Do you have happy good germs in your gut?
KEFIR
NATTO
SAUERKRAUT
KIMCHI
YOGURT

I hope it was fun to learn about us!

References

2. "The Good Gut: Taking Control of your Weight, Your Mood, and Your Long-term Health" by Justin Sonnenburg, Erica Sonnenburg, pg 4.

3. Stephen, A. M.; Cummings, J. H. (1980). "The Microbial Contribution to Human Faecal Mass". J. Med. Microbiology. 13 (1): 45–56.

4. "The Good Gut: Taking Control of your Weight, Your Mood, and Your Long-term Health" by Justin Sonnenburg, Erica Sonnenburg, pg 1-4.

7. Microbiome data collection project in US and UK, paused during Covid pandemic. https://microsetta.ucsd.edu

9. **Immune:** Wastyk HC, et. al. Gut-microbiota-targeted diets modulate human immune status. Cell. 2021 Aug 5;184(16):4137-4153.

Obesity: Le Chatelier E, et al. Richness of human gut microbiome correlates with metabolic markers. Nature. 2013 Aug 29; 500(7464):541-6.

Anxiety: Tillisch K, et al. Consumption of fermented milk product with probiotic modulates brain activity. Gastroenterology. 2013 Jun;144(7):1394-401, 1401.e1-4.

Lower Risk: "The Good Gut: Taking Control of your Weight, Your Mood, and Your Long-term Health" by Justin Sonnenburg, Erica Sonnenburg, pg 218.

10. Hensgens MP, et al. Time interval of increased risk for Clostridium difficile infection after exposure to antibiotics. J Antimicrob Chemother. 2012 Mar;67(3):742-8.

11. **Mood:** Messaoudi M, et al. Assessment of psychotropic-like properties of a probiotic formulation (Lactobacillus helveticus R0052 and Bifidobacterium longum R0175) in rats and human subjects. Br J Nutr. 2011 Mar;105(5):755-64.

Sleep: Szentirmai, É., et al. Butyrate, a metabolite of intestinal bacteria, enhances sleep. Sci Rep 9, 7035 (2019).

Immune system: Schulthess J, et al. The Short Chain Fatty Acid Butyrate Imprints an Antimicrobial Program in Macrophages. Immunity. 2019 Feb 19;50(2):432-445.

Blood sugar: De Vadder, F., et al. (2014). Microbiota-Generated Metabolites Promote Metabolic Benefits via Gut-Brain Neural Circuits. Cell 156, 84-96.

Suppress inflammation: Singh, N., et al. (2014). Activation of Gpr109a, Receptor for Niacin and the Commensal Metabolite Butyrate, Suppresses Colonic Inflammation and Carcinogenesis. Immunity 40, 128–139.

14. Chung WSF, et al. Prebiotic potential of pectin and pectin oligosaccharides to promote anti-inflammatory commensal bacteria in the human colon. FEMS Microbiol Ecol. 2017;93(11):fix127.

15. Kaczmarek, JL, et al. Broccoli consumption affects the human gastrointestinal microbiota, The Journal of Nutritional Biochemistry, Volume 63, 2019, Pages 27-34.

16. Ye, M., et al. Oatmeal induced gut microbiota alteration and its relationship with improved lipid profiles: a secondary analysis of a randomized clinical trial. Nutr Metab (Lond) 17, 85 (2020).

17. Creedon AC, et al. Nuts and their Effect on Gut Microbiota, Gut Function and Symptoms in Adults: A Systematic Review and Meta-Analysis of Randomised Controlled Trials. Nutrients. 2020 Aug 6;12(8):2347.

18. Abstract LB223: Beans to Enrich the Gut microbiome vs. Obesity's Negative Effects: First results from the BE GONE Trial in high-risk colorectal patients. Zhang, XT; et al. Cancer Res (2021) 81 (13_Supplement): LB223.

19. Tzounis X, et al. Prebiotic evaluation of cocoa-derived flavanols in healthy humans by using a randomized, controlled, double-blind, crossover intervention study. Am J Clin Nutr. 2011 Jan;93(1):62-72.

22. Wastyk HC, et al. Gut-microbiota-targeted diets modulate human immune status. Cell. 2021 Aug 5;184(16):4137-4153.

23. **Sauerkraut:** Lu, Z., et al. "Bacteriophage ecology in commercial sauerkraut fermentations." Applied and environmental microbiology vol. 69,6 (2003): 3192-202.

Kimchi, Allergies: Kwon, YS et al. "Relationship Between Plant Food (Fruits, Vegetables, and Kimchi) Consumption and the Prevalence of Rhinitis Among Korean Adults: Based on the 2011 and 2012 Korea National Health and Nutrition Examination Survey Data." Journal of medicinal food vol. 19,12 (2016): 1130-1140.

Kimchi, Asthma: Kim H, et al. Association between kimchi intake and asthma in Korean adults: the Fourth and Fifth Korea National Health and Nutrition Examination Survey (2007–2011). J Med Food 2014;17(1):172–178.

Kimchi, Eczema: Kim HJ, et al. Kimchi intake and atopic dematitis in Korean aged 19-49 years: the Korean National Health and Nutrition Examination Survey 2010–2012. Asia Pac J Clin Nutr 2017;26(5):914–922.

24. **Yogurt, Heart disease:** Ivey, K.L., et al. Association between yogurt, milk, and cheese consumption and common carotid artery intima-media thickness and cardiovascular disease risk factors in elderly women. Am J Clin Nutr. 2011 Jul;94(1):234-9.

Yogurt, Diabetes: Margolis, K.L., et al. A diet high in low-fat dairy products lowers diabetes risk in postmenopausal women. J Nutr. 2011 Nov;141(11):1969-74.

Kefir: Ostadrahimi, A., et al. Effect of probiotic fermented milk (kefir) on glycemic control and lipid profile in type 2 diabetic patients: a randomized double-blind placebo-controlled clinical trial. Iran J Public Health. 2015 Feb;44(2):228-37.

25. **Vitamin K2:** Mandatori D., et al. The Dual Role of Vitamin K2 in "Bone-Vascular Crosstalk": Opposite Effects on Bone Loss and Vascular Calcification. Nutrients. 2021 Apr 7;13(4):1222.

Virus: Lefevre, M., et al. Probiotic strain Bacillus subtilis CU1 stimulates immune system of elderly during common infectious disease period: a randomized, double-blind placebo-controlled study. Immun Ageing 12, 24 (2015).

26. "The Good Gut: Taking Control of your Weight, Your Mood, and Your Long-term Health" by Justin Sonnenburg, Erica Sonnenburg, pg 103.

About the Author

Rene Chee was inquisitive from a young age, always asking questions. As she grew older, this curiosity led her to pursue a Ph.D. in biology and conduct research as a postdoctoral researcher. However, during this time she was diagnosed with a rare cancer that could not be cured with conventional medicine. In her journey to survive, Rene learned firsthand the crucial importance of making science-based decisions. Now, 15 years later in 2023, she is healthy and continues to be passionate about using science to make decisions for everyday life.

Rene is the co-author of “Curing Cancer with Immunotherapy”, a book that provides a layperson's guide to cancer immunotherapy. She holds an M.S. in Biological Sciences from Stanford University, a Ph.D. in Biological Sciences from the University of California, San Diego, and was a postdoctoral scholar at the Stanford University School of Medicine.

Made in the USA
Las Vegas, NV
10 November 2026

57843104R00021